THIS BOOK BELONGS TO:

CONTACT INFORMATION	
NAME:	
ADDRESS:	
PHONE:	

START / END DATES

___ / ___ / ___   TO   ___ / ___ / ___

DEDICATION

This book is dedicated to all the energetic and hard working Farmers out there who want to keep accurate information & notes about their farm.

You are my inspiration for producing books and I'm honored to be a part of keeping all of your Farming notes and records organized.

This journal notebook will help you record your details about your farm or ranch organized.

Thoughtfully put together with these sections to record: Business Details, Livestock Record, Equipment Maintenance & Repair, Farm Expenses & Income, & Notes.

HOW TO USE THIS BOOK

The purpose of this book is to keep all of your Farming notes all in one place. It will help keep you organized.

This Farm Log Journal will allow you to accurately document every detail about your Farm. It's a great way to chart your course as you work your farm.

Here are examples of the prompts for you to fill in and write about your experience in this book:

1. Contact Page - Write your Name, Address, Phone Number, Start & End Dates of this Record.

2. Business Details - Write Business Name, Address, Email, Fax, Phone, Website, Logbook Details, & Notes.

3. Livestock Record - Date, Number, Type, Start & End: Quantity, Average Weight, Value, Balance.

4. Equipment Maintenance & Repair - Date, Month, Equipment, Inspection, Maintenance, Repair, Services Required, Date of Service or Repair

5. Farm Expenses - Date, Expenses, Cost, Remarks

6. Farm Income - Date, Source, Description, Method of Payment, Amount

7. Notes - Any other important information you wish to record such as farms inventory, garden plan, goat tracker, monthly notes, accounting notes, financial note & records, etc.

Enjoy!

BUSINESS DETAILS

BUSINESS NAME:	
ADDRESS:	
EMAIL ADDRESS:	PHONE NUMBER:
FAX NUMBER:	WEBSITE:

LOGBOOK DETAILS

CONTINUED FROM LOGBOOK:	CONTINUED TO LOG BOOK:
DATE LOG STARTED:	DATE LOG ENDED:

NOTES

LIVESTOCK RECORD

DATE:		START DATE:			SEND OF YEAR:			
NO	TYPE OF LIVESTOCK	QTY	AVG WEIGHT	VALUE	QTY	AVG WEIGHT	VALUE	BALANCE

EQUIPMENT MAINTENANCE & REPAIR

MONTH:					
DATE	EQUIPMENT	INSPECTION / MAINTENANCE / REPAIR / SERVICES REQUIRED	SERVICE / REPAIR DATE	INITIALS	REMARKS

FARM EXPENSES

MONTH:			
DATE	EXPENSES	COST	REMARKS

FARM INCOME

DATE	SOURCE	DESCRIPTION	METHOD OF PAYMENT	AMOUNT
			TOTAL:	

NOTES

BUSINESS DETAILS

BUSINESS NAME:	
ADDRESS:	
EMAIL ADDRESS:	PHONE NUMBER:
FAX NUMBER:	WEBSITE:

LOGBOOK DETAILS

CONTINUED FROM LOGBOOK:	CONTINUED TO LOG BOOK:
DATE LOG STARTED:	DATE LOG ENDED:

NOTES

LIVESTOCK RECORD

DATE: START DATE: SEND OF YEAR:

NO	TYPE OF LIVESTOCK	QTY	AVG WEIGHT	VALUE	QTY	AVG WEIGHT	VALUE	BALANCE

EQUIPMENT MAINTENANCE & REPAIR

MONTH:	

DATE	EQUIPMENT	INSPECTION / MAINTENANCE / REPAIR / SERVICES REQUIRED	SERVICE / REPAIR DATE	INITIALS	REMARKS

FARM EXPENSES

MONTH:			
DATE	EXPENSES	COST	REMARKS

FARM INCOME

DATE	SOURCE	DESCRIPTION	METHOD OF PAYMENT	AMOUNT
		TOTAL:	

NOTES

BUSINESS DETAILS

BUSINESS NAME:	
ADDRESS:	

EMAIL ADDRESS:	PHONE NUMBER:
FAX NUMBER:	WEBSITE:

LOGBOOK DETAILS

CONTINUED FROM LOGBOOK:	CONTINUED TO LOG BOOK:
DATE LOG STARTED:	DATE LOG ENDED:

NOTES

LIVESTOCK RECORD

DATE:		START DATE:			SEND OF YEAR:			
NO	TYPE OF LIVESTOCK	QTY	AVG WEIGHT	VALUE	QTY	AVG WEIGHT	VALUE	BALANCE

EQUIPMENT MAINTENANCE & REPAIR

MONTH:					
DATE	EQUIPMENT	INSPECTION / MAINTENANCE / REPAIR / SERVICES REQUIRED	SERVICE / REPAIR DATE	INITIALS	REMARKS

FARM EXPENSES

MONTH:			
DATE	EXPENSES	COST	REMARKS

FARM INCOME

DATE	SOURCE	DESCRIPTION	METHOD OF PAYMENT	AMOUNT
			TOTAL:	

NOTES

BUSINESS DETAILS

BUSINESS NAME:	
ADDRESS:	
EMAIL ADDRESS:	PHONE NUMBER:
FAX NUMBER:	WEBSITE:

LOGBOOK DETAILS

CONTINUED FROM LOGBOOK:	CONTINUED TO LOG BOOK:
DATE LOG STARTED:	DATE LOG ENDED:

NOTES

LIVESTOCK RECORD

DATE:		START DATE:			SEND OF YEAR:			
NO	TYPE OF LIVESTOCK	QTY	AVG WEIGHT	VALUE	QTY	AVG WEIGHT	VALUE	BALANCE

EQUIPMENT MAINTENANCE & REPAIR

MONTH:					
DATE	EQUIPMENT	INSPECTION / MAINTENANCE / REPAIR / SERVICES REQUIRED	SERVICE / REPAIR DATE	INITIALS	REMARKS

FARM EXPENSES

MONTH:			
DATE	EXPENSES	COST	REMARKS

FARM INCOME

DATE	SOURCE	DESCRIPTION	METHOD OF PAYMENT	AMOUNT
			TOTAL:	

NOTES

BUSINESS DETAILS

BUSINESS NAME:	
ADDRESS:	

EMAIL ADDRESS:	PHONE NUMBER:
FAX NUMBER:	WEBSITE:

LOGBOOK DETAILS

CONTINUED FROM LOGBOOK:	CONTINUED TO LOG BOOK:
DATE LOG STARTED:	DATE LOG ENDED:

NOTES

LIVESTOCK RECORD

DATE: **START DATE:** **SEND OF YEAR:**

NO	TYPE OF LIVESTOCK	QTY	AVG WEIGHT	VALUE	QTY	AVG WEIGHT	VALUE	BALANCE

EQUIPMENT MAINTENANCE & REPAIR

MONTH:					
DATE	EQUIPMENT	INSPECTION / MAINTENANCE / REPAIR / SERVICES REQUIRED	SERVICE / REPAIR DATE	INITIALS	REMARKS

FARM EXPENSES

MONTH:			
DATE	EXPENSES	COST	REMARKS

FARM INCOME

DATE	SOURCE	DESCRIPTION	METHOD OF PAYMENT	AMOUNT
			TOTAL:	

NOTES

BUSINESS DETAILS

BUSINESS NAME:	
ADDRESS:	

EMAIL ADDRESS:	PHONE NUMBER:
FAX NUMBER:	WEBSITE:

LOGBOOK DETAILS

CONTINUED FROM LOGBOOK:	CONTINUED TO LOG BOOK:
DATE LOG STARTED:	DATE LOG ENDED:

NOTES

LIVESTOCK RECORD

DATE:		START DATE:			SEND OF YEAR:			
NO	TYPE OF LIVESTOCK	QTY	AVG WEIGHT	VALUE	QTY	AVG WEIGHT	VALUE	BALANCE

EQUIPMENT MAINTENANCE & REPAIR

MONTH:					
DATE	EQUIPMENT	INSPECTION / MAINTENANCE / REPAIR / SERVICES REQUIRED	SERVICE / REPAIR DATE	INITIALS	REMARKS

FARM EXPENSES

MONTH:			
DATE	EXPENSES	COST	REMARKS

FARM INCOME

DATE	SOURCE	DESCRIPTION	METHOD OF PAYMENT	AMOUNT
			TOTAL:	

NOTES

BUSINESS DETAILS

BUSINESS NAME:	
ADDRESS:	

EMAIL ADDRESS:	PHONE NUMBER:
FAX NUMBER:	WEBSITE:

LOGBOOK DETAILS

CONTINUED FROM LOGBOOK:	CONTINUED TO LOG BOOK:
DATE LOG STARTED:	DATE LOG ENDED:

NOTES

LIVESTOCK RECORD

DATE: START DATE: SEND OF YEAR:

NO	TYPE OF LIVESTOCK	QTY	AVG WEIGHT	VALUE	QTY	AVG WEIGHT	VALUE	BALANCE

EQUIPMENT MAINTENANCE & REPAIR

MONTH:

DATE	EQUIPMENT	INSPECTION / MAINTENANCE / REPAIR / SERVICES REQUIRED	SERVICE / REPAIR DATE	INITIALS	REMARKS

FARM EXPENSES

MONTH:			
DATE	EXPENSES	COST	REMARKS

FARM INCOME

DATE	SOURCE	DESCRIPTION	METHOD OF PAYMENT	AMOUNT
			TOTAL:	

NOTES

BUSINESS DETAILS

BUSINESS NAME:	
ADDRESS:	

EMAIL ADDRESS:	PHONE NUMBER:
FAX NUMBER:	WEBSITE:

LOGBOOK DETAILS

CONTINUED FROM LOGBOOK:	CONTINUED TO LOG BOOK:
DATE LOG STARTED:	DATE LOG ENDED:

NOTES

LIVESTOCK RECORD

DATE:		START DATE:			SEND OF YEAR:			
NO	TYPE OF LIVESTOCK	QTY	AVG WEIGHT	VALUE	QTY	AVG WEIGHT	VALUE	BALANCE

EQUIPMENT MAINTENANCE & REPAIR

MONTH:					
DATE	EQUIPMENT	INSPECTION / MAINTENANCE / REPAIR / SERVICES REQUIRED	SERVICE / REPAIR DATE	INITIALS	REMARKS

FARM EXPENSES

MONTH:			
DATE	EXPENSES	COST	REMARKS

FARM INCOME

DATE	SOURCE	DESCRIPTION	METHOD OF PAYMENT	AMOUNT
			TOTAL:	

NOTES

BUSINESS DETAILS

BUSINESS NAME:	
ADDRESS:	
EMAIL ADDRESS:	PHONE NUMBER:
FAX NUMBER:	WEBSITE:

LOGBOOK DETAILS

CONTINUED FROM LOGBOOK:	CONTINUED TO LOG BOOK:
DATE LOG STARTED:	DATE LOG ENDED:

NOTES

LIVESTOCK RECORD

DATE: **START DATE:** **SEND OF YEAR:**

NO	TYPE OF LIVESTOCK	QTY	AVG WEIGHT	VALUE	QTY	AVG WEIGHT	VALUE	BALANCE

EQUIPMENT MAINTENANCE & REPAIR

MONTH:					
DATE	EQUIPMENT	INSPECTION / MAINTENANCE / REPAIR / SERVICES REQUIRED	SERVICE / REPAIR DATE	INITIALS	REMARKS

FARM EXPENSES

MONTH:			
DATE	EXPENSES	COST	REMARKS

FARM INCOME

DATE	SOURCE	DESCRIPTION	METHOD OF PAYMENT	AMOUNT
			TOTAL:	

NOTES

BUSINESS DETAILS

BUSINESS NAME:	
ADDRESS:	
EMAIL ADDRESS:	PHONE NUMBER:
FAX NUMBER:	WEBSITE:

LOGBOOK DETAILS

CONTINUED FROM LOGBOOK:	CONTINUED TO LOG BOOK:
DATE LOG STARTED:	DATE LOG ENDED:

NOTES

LIVESTOCK RECORD

DATE:		START DATE:			SEND OF YEAR:			
NO	TYPE OF LIVESTOCK	QTY	AVG WEIGHT	VALUE	QTY	AVG WEIGHT	VALUE	BALANCE

EQUIPMENT MAINTENANCE & REPAIR

MONTH:					
DATE	EQUIPMENT	INSPECTION / MAINTENANCE / REPAIR / SERVICES REQUIRED	SERVICE / REPAIR DATE	INITIALS	REMARKS

FARM EXPENSES

MONTH:			
DATE	EXPENSES	COST	REMARKS

FARM INCOME

DATE	SOURCE	DESCRIPTION	METHOD OF PAYMENT	AMOUNT
			TOTAL:	

NOTES

BUSINESS DETAILS

BUSINESS NAME:	
ADDRESS:	

EMAIL ADDRESS:	PHONE NUMBER:
FAX NUMBER:	WEBSITE:

LOGBOOK DETAILS

CONTINUED FROM LOGBOOK:	CONTINUED TO LOG BOOK:
DATE LOG STARTED:	DATE LOG ENDED:

NOTES

LIVESTOCK RECORD

DATE:

START DATE:

SEND OF YEAR:

NO	TYPE OF LIVESTOCK	QTY	AVG WEIGHT	VALUE	QTY	AVG WEIGHT	VALUE	BALANCE

EQUIPMENT MAINTENANCE & REPAIR

MONTH:					
DATE	EQUIPMENT	INSPECTION / MAINTENANCE / REPAIR / SERVICES REQUIRED	SERVICE / REPAIR DATE	INITIALS	REMARKS

FARM EXPENSES

MONTH:			
DATE	EXPENSES	COST	REMARKS

FARM INCOME

DATE	SOURCE	DESCRIPTION	METHOD OF PAYMENT	AMOUNT
			TOTAL:	

NOTES

BUSINESS DETAILS

BUSINESS NAME:	
ADDRESS:	

EMAIL ADDRESS:	PHONE NUMBER:
FAX NUMBER:	WEBSITE:

LOGBOOK DETAILS

CONTINUED FROM LOGBOOK:	CONTINUED TO LOG BOOK:
DATE LOG STARTED:	DATE LOG ENDED:

NOTES

LIVESTOCK RECORD

DATE:		START DATE:			SEND OF YEAR:			
NO	TYPE OF LIVESTOCK	QTY	AVG WEIGHT	VALUE	QTY	AVG WEIGHT	VALUE	BALANCE

EQUIPMENT MAINTENANCE & REPAIR

MONTH:

DATE	EQUIPMENT	INSPECTION / MAINTENANCE / REPAIR / SERVICES REQUIRED	SERVICE / REPAIR DATE	INITIALS	REMARKS

FARM EXPENSES

MONTH:			
DATE	EXPENSES	COST	REMARKS

FARM INCOME

DATE	SOURCE	DESCRIPTION	METHOD OF PAYMENT	AMOUNT
			TOTAL:	

NOTES

BUSINESS DETAILS

BUSINESS NAME:	
ADDRESS:	
EMAIL ADDRESS:	PHONE NUMBER:
FAX NUMBER:	WEBSITE:

LOGBOOK DETAILS

CONTINUED FROM LOGBOOK:	CONTINUED TO LOG BOOK:
DATE LOG STARTED:	DATE LOG ENDED:

NOTES

LIVESTOCK RECORD

DATE:		START DATE:			SEND OF YEAR:			
NO	TYPE OF LIVESTOCK	QTY	AVG WEIGHT	VALUE	QTY	AVG WEIGHT	VALUE	BALANCE

EQUIPMENT MAINTENANCE & REPAIR

MONTH:	

DATE	EQUIPMENT	INSPECTION / MAINTENANCE / REPAIR / SERVICES REQUIRED	SERVICE / REPAIR DATE	INITIALS	REMARKS

FARM EXPENSES

MONTH:			
DATE	EXPENSES	COST	REMARKS

FARM INCOME

DATE	SOURCE	DESCRIPTION	METHOD OF PAYMENT	AMOUNT
			TOTAL:	

NOTES

BUSINESS DETAILS

BUSINESS NAME:	
ADDRESS:	

EMAIL ADDRESS:	PHONE NUMBER:
FAX NUMBER:	WEBSITE:

LOGBOOK DETAILS

CONTINUED FROM LOGBOOK:	CONTINUED TO LOG BOOK:
DATE LOG STARTED:	DATE LOG ENDED:

NOTES

LIVESTOCK RECORD

DATE: START DATE: SEND OF YEAR:

NO	TYPE OF LIVESTOCK	QTY	AVG WEIGHT	VALUE	QTY	AVG WEIGHT	VALUE	BALANCE

EQUIPMENT MAINTENANCE & REPAIR

MONTH:	

DATE	EQUIPMENT	INSPECTION / MAINTENANCE / REPAIR / SERVICES REQUIRED	SERVICE / REPAIR DATE	INITIALS	REMARKS

FARM EXPENSES

MONTH:			
DATE	EXPENSES	COST	REMARKS

FARM INCOME

DATE	SOURCE	DESCRIPTION	METHOD OF PAYMENT	AMOUNT
			TOTAL:	

NOTES

BUSINESS DETAILS

BUSINESS NAME:	
ADDRESS:	

EMAIL ADDRESS:	PHONE NUMBER:
FAX NUMBER:	WEBSITE:

LOGBOOK DETAILS

CONTINUED FROM LOGBOOK:	CONTINUED TO LOG BOOK:
DATE LOG STARTED:	DATE LOG ENDED:

NOTES

LIVESTOCK RECORD

DATE:		START DATE:			SEND OF YEAR:			
NO	TYPE OF LIVESTOCK	QTY	AVG WEIGHT	VALUE	QTY	AVG WEIGHT	VALUE	BALANCE

EQUIPMENT MAINTENANCE & REPAIR

MONTH:					
DATE	EQUIPMENT	INSPECTION / MAINTENANCE / REPAIR / SERVICES REQUIRED	SERVICE / REPAIR DATE	INITIALS	REMARKS

FARM EXPENSES

MONTH:			
DATE	EXPENSES	COST	REMARKS

FARM INCOME

DATE	SOURCE	DESCRIPTION	METHOD OF PAYMENT	AMOUNT
			TOTAL:	

NOTES

BUSINESS DETAILS

BUSINESS NAME:	
ADDRESS:	

EMAIL ADDRESS:	PHONE NUMBER:
FAX NUMBER:	WEBSITE:

LOGBOOK DETAILS

CONTINUED FROM LOGBOOK:	CONTINUED TO LOG BOOK:
DATE LOG STARTED:	DATE LOG ENDED:

NOTES

LIVESTOCK RECORD

DATE: **START DATE:** **SEND OF YEAR:**

NO	TYPE OF LIVESTOCK	QTY	AVG WEIGHT	VALUE	QTY	AVG WEIGHT	VALUE	BALANCE

EQUIPMENT MAINTENANCE & REPAIR

MONTH:	

DATE	EQUIPMENT	INSPECTION / MAINTENANCE / REPAIR / SERVICES REQUIRED	SERVICE / REPAIR DATE	INITIALS	REMARKS

FARM EXPENSES

MONTH:			
DATE	EXPENSES	COST	REMARKS

FARM INCOME

DATE	SOURCE	DESCRIPTION	METHOD OF PAYMENT	AMOUNT
			TOTAL:	

NOTES

BUSINESS DETAILS

BUSINESS NAME:	
ADDRESS:	

EMAIL ADDRESS:	PHONE NUMBER:
FAX NUMBER:	WEBSITE:

LOGBOOK DETAILS

CONTINUED FROM LOGBOOK:	CONTINUED TO LOG BOOK:
DATE LOG STARTED:	DATE LOG ENDED:

NOTES

LIVESTOCK RECORD

DATE:		START DATE:			SEND OF YEAR:			
NO	TYPE OF LIVESTOCK	QTY	AVG WEIGHT	VALUE	QTY	AVG WEIGHT	VALUE	BALANCE

EQUIPMENT MAINTENANCE & REPAIR

MONTH:	

DATE	EQUIPMENT	INSPECTION / MAINTENANCE / REPAIR / SERVICES REQUIRED	SERVICE / REPAIR DATE	INITIALS	REMARKS

FARM EXPENSES

MONTH:			
DATE	EXPENSES	COST	REMARKS

FARM INCOME

DATE	SOURCE	DESCRIPTION	METHOD OF PAYMENT	AMOUNT
			TOTAL:	

NOTES

BUSINESS DETAILS

BUSINESS NAME:	
ADDRESS:	

EMAIL ADDRESS:	PHONE NUMBER:
FAX NUMBER:	WEBSITE:

LOGBOOK DETAILS

CONTINUED FROM LOGBOOK:	CONTINUED TO LOG BOOK:
DATE LOG STARTED:	DATE LOG ENDED:

NOTES

LIVESTOCK RECORD

DATE:		START DATE:			SEND OF YEAR:			
NO	TYPE OF LIVESTOCK	QTY	AVG WEIGHT	VALUE	QTY	AVG WEIGHT	VALUE	BALANCE

EQUIPMENT MAINTENANCE & REPAIR

MONTH:

DATE	EQUIPMENT	INSPECTION / MAINTENANCE / REPAIR / SERVICES REQUIRED	SERVICE / REPAIR DATE	INITIALS	REMARKS

FARM EXPENSES

MONTH:

DATE	EXPENSES	COST	REMARKS

FARM INCOME

DATE	SOURCE	DESCRIPTION	METHOD OF PAYMENT	AMOUNT
			TOTAL:	

NOTES

BUSINESS DETAILS

BUSINESS NAME:	
ADDRESS:	
EMAIL ADDRESS:	PHONE NUMBER:
FAX NUMBER:	WEBSITE:

LOGBOOK DETAILS

CONTINUED FROM LOGBOOK:	CONTINUED TO LOG BOOK:
DATE LOG STARTED:	DATE LOG ENDED:

NOTES

LIVESTOCK RECORD

DATE:		START DATE:			SEND OF YEAR:			
NO	TYPE OF LIVESTOCK	QTY	AVG WEIGHT	VALUE	QTY	AVG WEIGHT	VALUE	BALANCE

EQUIPMENT MAINTENANCE & REPAIR

MONTH:					
DATE	EQUIPMENT	INSPECTION / MAINTENANCE / REPAIR / SERVICES REQUIRED	SERVICE / REPAIR DATE	INITIALS	REMARKS

FARM EXPENSES

MONTH:			
DATE	EXPENSES	COST	REMARKS

FARM INCOME

DATE	SOURCE	DESCRIPTION	METHOD OF PAYMENT	AMOUNT
			TOTAL:	

NOTES

BUSINESS DETAILS

BUSINESS NAME:	
ADDRESS:	

EMAIL ADDRESS:	PHONE NUMBER:
FAX NUMBER:	WEBSITE:

LOGBOOK DETAILS

CONTINUED FROM LOGBOOK:	CONTINUED TO LOG BOOK:
DATE LOG STARTED:	DATE LOG ENDED:

NOTES

LIVESTOCK RECORD

DATE:		START DATE:			SEND OF YEAR:			
NO	TYPE OF LIVESTOCK	QTY	AVG WEIGHT	VALUE	QTY	AVG WEIGHT	VALUE	BALANCE

EQUIPMENT MAINTENANCE & REPAIR

MONTH:

DATE	EQUIPMENT	INSPECTION / MAINTENANCE / REPAIR / SERVICES REQUIRED	SERVICE / REPAIR DATE	INITIALS	REMARKS

FARM EXPENSES

MONTH:			
DATE	EXPENSES	COST	REMARKS

FARM INCOME

DATE	SOURCE	DESCRIPTION	METHOD OF PAYMENT	AMOUNT
			TOTAL:	

NOTES

BUSINESS DETAILS

BUSINESS NAME:	
ADDRESS:	

EMAIL ADDRESS:	PHONE NUMBER:
FAX NUMBER:	WEBSITE:

LOGBOOK DETAILS

CONTINUED FROM LOGBOOK:	CONTINUED TO LOG BOOK:
DATE LOG STARTED:	DATE LOG ENDED:

NOTES

LIVESTOCK RECORD

DATE:		START DATE:			SEND OF YEAR:			
NO	TYPE OF LIVESTOCK	QTY	AVG WEIGHT	VALUE	QTY	AVG WEIGHT	VALUE	BALANCE

EQUIPMENT MAINTENANCE & REPAIR

MONTH:	

DATE	EQUIPMENT	INSPECTION / MAINTENANCE / REPAIR / SERVICES REQUIRED	SERVICE / REPAIR DATE	INITIALS	REMARKS

FARM EXPENSES

MONTH:

DATE	EXPENSES	COST	REMARKS

FARM INCOME

DATE	SOURCE	DESCRIPTION	METHOD OF PAYMENT	AMOUNT
			TOTAL:	

NOTES

BUSINESS DETAILS

BUSINESS NAME:	
ADDRESS:	
EMAIL ADDRESS:	PHONE NUMBER:
FAX NUMBER:	WEBSITE:

LOGBOOK DETAILS

CONTINUED FROM LOGBOOK:	CONTINUED TO LOG BOOK:
DATE LOG STARTED:	DATE LOG ENDED:

NOTES

LIVESTOCK RECORD

DATE:		START DATE:			SEND OF YEAR:				
NO	TYPE OF LIVESTOCK	QTY	AVG WEIGHT	VALUE	QTY	AVG WEIGHT	VALUE	BALANCE	

EQUIPMENT MAINTENANCE & REPAIR

MONTH:					
DATE	EQUIPMENT	INSPECTION / MAINTENANCE / REPAIR / SERVICES REQUIRED	SERVICE / REPAIR DATE	INITIALS	REMARKS

FARM EXPENSES

MONTH:			
DATE	EXPENSES	COST	REMARKS

FARM INCOME

DATE	SOURCE	DESCRIPTION	METHOD OF PAYMENT	AMOUNT
			TOTAL:	

NOTES

BUSINESS DETAILS

BUSINESS NAME:	
ADDRESS:	
EMAIL ADDRESS:	PHONE NUMBER:
FAX NUMBER:	WEBSITE:

LOGBOOK DETAILS

CONTINUED FROM LOGBOOK:	CONTINUED TO LOG BOOK:
DATE LOG STARTED:	DATE LOG ENDED:

NOTES

LIVESTOCK RECORD

DATE:		START DATE:			SEND OF YEAR:			
NO	TYPE OF LIVESTOCK	QTY	AVG WEIGHT	VALUE	QTY	AVG WEIGHT	VALUE	BALANCE

EQUIPMENT MAINTENANCE & REPAIR

MONTH:					
DATE	EQUIPMENT	INSPECTION / MAINTENANCE / REPAIR / SERVICES REQUIRED	SERVICE / REPAIR DATE	INITIALS	REMARKS

FARM EXPENSES

MONTH:			
DATE	EXPENSES	COST	REMARKS

FARM INCOME

DATE	SOURCE	DESCRIPTION	METHOD OF PAYMENT	AMOUNT
			TOTAL:	

NOTES

BUSINESS DETAILS

BUSINESS NAME:	
ADDRESS:	
EMAIL ADDRESS:	PHONE NUMBER:
FAX NUMBER:	WEBSITE:

LOGBOOK DETAILS

CONTINUED FROM LOGBOOK:	CONTINUED TO LOG BOOK:
DATE LOG STARTED:	DATE LOG ENDED:

NOTES

LIVESTOCK RECORD

DATE:		START DATE:			SEND OF YEAR:			
NO	TYPE OF LIVESTOCK	QTY	AVG WEIGHT	VALUE	QTY	AVG WEIGHT	VALUE	BALANCE

EQUIPMENT MAINTENANCE & REPAIR

MONTH:					
DATE	EQUIPMENT	INSPECTION / MAINTENANCE / REPAIR / SERVICES REQUIRED	SERVICE / REPAIR DATE	INITIALS	REMARKS

FARM EXPENSES

MONTH:			
DATE	EXPENSES	COST	REMARKS

FARM INCOME

DATE	SOURCE	DESCRIPTION	METHOD OF PAYMENT	AMOUNT
			TOTAL:	

www.ingramcontent.com/pod-product-compliance
Lightning Source LLC
Chambersburg PA
CBHW081229080526
44587CB00022B/3876